W9-CFX-521

A New True Book

SPACELAB

By Dennis B. Fradin

629.44
FRA

CHILDRENS PRESS ™

CHICAGO

CLEVE. HTS.- UNIV. HTS.
PUBLIC SCHOOLS
MEDIA CENTERS

NOBLE SCHOOL
MEDIA CENTER

For David Polster

Spacelab was built by the
European Space Agency.

PHOTO CREDITS

National Aeronautics and Space
Administration:

John F. Kennedy Space Center— cover, 4, 8, 14, 15
23, 25, 29, 32, 36, 41, 43 (left)

Marshall Space Flight Center—13 (7 photos),
19, 22, 27 (left), 30, 34, 39 (2 photos), 42, 43
(right), 44

ESA (European Space Agency)—2, 7, 11, 16 (2
photos), 17, 20, 22, 27 (right), 28

Library of Congress Cataloging in Publication Data

Fradin, Dennis B.
  Spacelab.

  (A New true book)
  Summary: An introduction to Spacelab, a reusable
space laboratory for studying many scientific areas,
first sent aloft in 1983.
  1. Spacelab Project— Juvenile literature.
[1. Spacelab Project] I. . Title.
TL797.F73 1983   629.44′5   84-12702
ISBN 0-516-01930-9   AACR2

Copyright© 1984 by Regensteiner Publishing Enterprises, Inc.
All rights reserved. Published simultaneously in Canada.
Printed in the United States of America.
1 2 3 4 5 6 7 8 9 10 R 93 92 91 90 89 88 87 86 85 84

8502958 N

TABLE OF CONTENTS

SPACELAB LIFTS OFF!

On the morning of November 28, 1983, thousands of people gathered at Cape Canaveral, Florida. They had come to watch the space shuttle *Columbia* carry six men and a laboratory called Spacelab into outer space.

Twice earlier *Columbia* had been ready to take Spacelab into space. Equipment failures had scrapped both flights.

Now, thousands of onlookers, hundreds of scientists who had planned the mission, and the six-man crew hoped the launch would go smoothly.

At 11:00 on that Monday morning, *Columbia* blasted off. "Looks pretty good!" Commander John Young

This artist's drawing shows how Spacelab fits in the cargo bay of the *Columbia.*

reported. Less than an hour later, *Columbia,* with Spacelab inside, was in orbit 155 miles above the earth. The crew began to prepare for nine days of experiments aboard Spacelab.

The astronauts who lived in Skylab proved that people could live and work in space for long periods of time.

WHAT IS SPACELAB?

In May of 1973 the
United States sent the first
space laboratory into orbit.
Its name was Skylab.
During the next eight
months three crews
rocketed up to live and
work in Skylab.

The Skylab astronauts
wanted to see how well
people can function in
space for long periods.
They found they were able

to live and work there
quite well. The Skylab
crew did experiments.
They studied the earth and
the sun.

After the third crew left,
Skylab was abandoned.
Finally Skylab fell from the
sky. The $2.5 billion craft
burned up in the earth's
atmosphere.

Because of Skylab, scientists
planned another space
laboratory, called Spacelab.

Cutaway view of the interior of Spacelab.
Because the lab has earthlike environment, the
scientists can work without space suits.

It would be better than
Skylab. It would be
reusable. It would be
carried by a shuttle (a
cargo ship), which could
also be reused.

The United States and thirteen other nations planned Spacelab and sent it up on its first mission.

Columbia, the shuttle, was built by the United States. *Columbia* combined the features of three kinds of spaceships. It could blast off like a rocket, orbit like a space station, and land on a runway like a glider.

(1) Shuttle's main engines and rocket boosters launch Spacelab.

(2) Rocket boosters separate and parachute to sea where they are picked up for reuse.

(3) The external fuel tank separates from the shuttle just before it enters its orbit. This tank breaks up in space and cannot be reused.

(4) The shuttle can use its power to adjust its path in orbit and to slow down its speed when it returns to earth.

(5) The space shuttle's cargo bay doors can be opened in space.

(6) A special heat shield protects the shuttle from burning up in the earth's atmosphere upon its reentry.

(7) The space shuttle glides to its landing.

The Spacelab was built in Bremen, West Germany.

Spacelab was built in West Germany by the eleven-nation European Space Agency. Spacelab wasn't a single laboratory. It was a set of parts that could be put together in different ways.

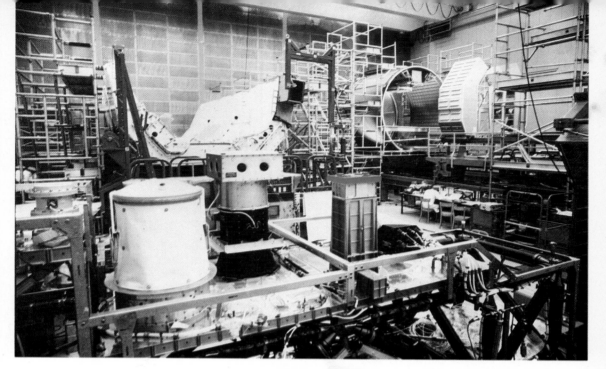

Spacelab has separate units that can be put together in different ways to suit the scientific needs of the mission.

For its first mission, Spacelab had two main parts. There was the twenty-three-foot-long laboratory. The laboratory had pressure, temperature, and air the same as earth.

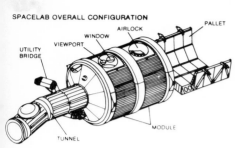

SPACELAB OVERALL CONFIGURATION

UTILITY BRIDGE

VIEWPORT

WINDOW

AIRLOCK

PALLET

MODULE

TUNNEL

Diagram of the different
parts of Spacelab (above).
Can you find these parts
in the photograph at right?
The pallet is at the
top of the picture.

The crew could work
without space suits. There
was also a U-shaped
section called the pallet at
the rear of the laboratory.

16

Close-up of the equipment on the U-shaped pallet

The pallet was exposed directly to outer space. It contained telescopes and other instruments used to study the heavens.

The seventy-two experiments on the first Spacelab mission were designed by scientists from fourteen countries—Austria, Belgium, Canada, Denmark, France, Italy, Japan, The Netherlands, Spain, Sweden, Switzerland, the United Kingdom, the United States, and West Germany. Some experiments were to be done by the crew. Others were to be run by the instruments on the pallet.

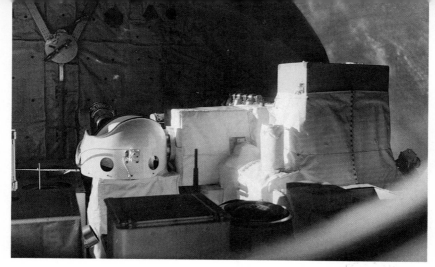

These scientific instruments (cameras and sensors) were carried on Spacelab's first flight.

Scientists hoped that the first Spacelab mission would answer some interesting questions: What happens to stars in their last years? What is the composition of the gases in the earth's upper atmosphere? Do plants need the pull of gravity to grow normally?

19

Spacelab crew from left to right: Back row: Dr. Byron K.
Lichtenberg (payload specialist), Dr. Ulf Merbold (payload
specialist)
Front row: Dr. Owen K. Garriott (mission specialist), Major
Brewster H. Shaw (pilot), Captain John W. Young (commander),
Dr. Robert A.R. Parker (mission specialist)

THE SPACELAB CREW

Six men made up the first Spacelab crew. Two of them—John W. Young and Brewster H. Shaw, Jr.—had the important job of piloting *Columbia.* Young, the crew commander, had been on five earlier space voyages, dating back to 1965. At the time of the Spacelab launch, Commander Young was the world's most

Payload specialists and mission specialists in lab
during a shift change. Crew members worked twelve-hour
shifts so work in Spacelab could go on twenty-four hours a day.

experienced space traveler.
Shaw, an Air Force test
pilot, was the pilot for the
mission.

There were two mission
specialists aboard—Dr. Owen
K. Garriott, who had been
on the second Skylab

Astronaut Robert Parker (left) and other crew members took part in biomedical tests that were conducted in weightless conditions of space.

mission in 1973, and Dr. Robert A.R. Parker, an astronomy professor. They were to keep Spacelab's systems working properly and also do some experiments.

Finally, there were two payload specialists—
Dr. Byron K. Lichtenberg, a researcher at the Massachusetts Institute of Technology, and Dr. Ulf Merbold, a scientist from West Germany. These scientists were responsible for the experiments.

The crew was the largest ever to go into space. It was the first crew launched by the

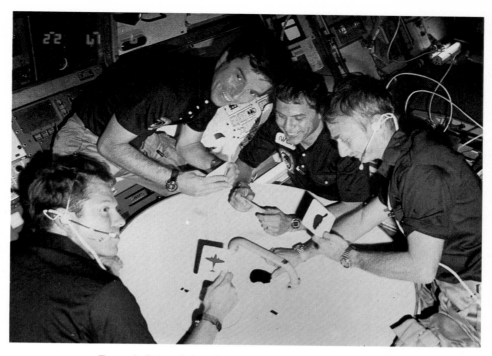

From left to right: Garriott, Parker, Merbold, and Lichtenberg

United States in which two members (Lichtenberg and Merbold) weren't astronauts. Merbold was also the first non-American to fly on a U.S. space mission.

Spacelab's first crew was unusual in one other way. "For the first time in spaceflight, the doctors outnumber the pilots four to two!" Commander Young said before launch. In the past, crews wanted to reach a place, explore it, and then return home. Such missions had needed spaceflight experts more than scientists. Spacelab, though, was dedicated to science. It required more scientists than pilots.

LIVING AND WORKING IN SPACE

Within several hours after launch, the Spacelab crew went to work on the experiments. To do work without stopping,

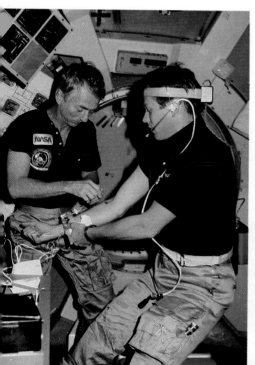

Dr. Owen Garriott takes blood from Byron Lichtenberg for an experiment. Dr. Ulf Merbold conducts an experiment in the science module.

Left to right: Ulf Merbold, Byron Lichtenberg, and Robert Parker

the crew was divided into
two shifts. Young, Parker,
and Merbold made up the
red shift. Shaw, Garriott,
and Lichtenberg were the
blue shift. Each shift
worked twelve hours and

Young and Merbold on the mid-deck of the *Columbia*.
Merbold's "headband" recorded how he adapted to the
environment in space. He wore it constantly.

rested the other twelve.

Columbia had living
quarters for the men
resting. There were cooking
and bathroom facilities,
plus three bunks for
sleeping. A narrow tunnel
connected *Columbia* to

Dr. Garriott comes through the tunnel that connected the Spacelab to the shuttle's crew compartment.

Spacelab so the men could go back and forth between the two.

The laboratory had many items found in earth

laboratories. There were test tubes, experiment racks, and scientific instruments.

Whether working or resting, the crew faced one unearthly condition. They were weightless. The spacecraft was high above the earth and moving fast (17,500 miles per hour). Gravity did not hold the men to the floor. Because of this, the laboratory and living quarters had special

Belts are used to hold astronauts in place when they sleep. Without them the astronauts would float off in the weightless conditions of space.

features. There were handrails that the crew used to keep from floating away. Test tubes were kept closed to keep liquids inside. The bunks had straps to keep the men from floating out of bed in their sleep.

WHAT SPACELAB DISCOVERED

The first Spacelab mission was a scientific success. Scientists learned about many things—from giant stars to tiny blood cells.

People in space had sometimes had problems with space motion sickness. The Spacelab crew learned that if space travelers keep their heads

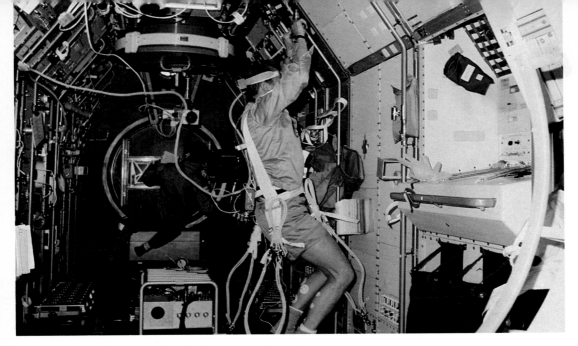

Byron Lichtenberg conducted an experiment
to find out more about motion sickness.

fairly still, they might avoid
motion sickness. They also
found that certain
medicines reduce the
symptoms of space motion
sickness. This information
should help future space
voyagers. The crew also

discovered that there are several changes in blood cells in weightless conditions.

Another experiment concerned plants. Living things go through cycles called circadian rhythms. For example, plants lift their leaves during daylight hours, then lower them at night. Are these rhythms caused by some force such as gravity, or are they built into the

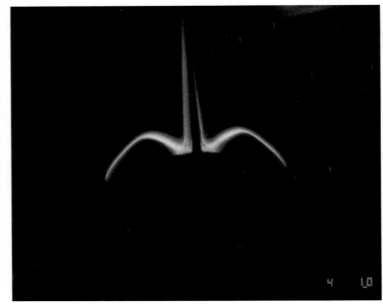

A camera took this photograph looking back at the tail of the space shuttle. Tests were done to see how materials reacted when they were exposed to conditions outside the earth's atmosphere.

organisms? Plants grown in Spacelab kept their daily patterns. This indicated that these rhythms are part of the plants' basic makeup.

Spacelab did several studies of the earth and other objects in space. The mission made the first upper-atmosphere measurements of a heavy form of hydrogen called deuterium. From this information scientists may be able to find out what elements formed our solar system. Spacelab also provided the best view ever of dying stars. In addition, Spacelab watched some hot, young stars being formed in our own Milky Way galaxy.

The Spacelab crew
made a new alloy—a
mixture of aluminum and
zinc. This strong,
lightweight alloy could be
used to build space
vehicles and airplanes.
Because of gravity, it is
difficult to make this alloy
on earth. One day it may
be mass-produced in
space and then shipped
back for use on earth.

Ulf Merbold (left), Byron Lichtenberg, and Owen
Garriott (right) took part in dozens of biology experiments.

TROUBLE ON
THE WAY HOME

The first Spacelab flight
was to last nine days. But
everything went so
smoothly that it was
extended for a tenth day.

The only serious problem occurred when *Columbia* began to return home. As *Columbia* fired its rockets, the jolt caused a computer to shut down. The computer was needed to land the ship.

Columbia had four other computers that could be used for landing. But, another blast of the rockets caused a second computer shutdown. It was

Portion of the Spacelab is pictured as the shuttle's camera photographed the People's Republic of Vietnam and the South China Sea from 155 miles above the earth.

feared that if *Columbia* tried to land, *all* the computers might break down. That would make it difficult to land safely.

There was a delay of more than seven hours.

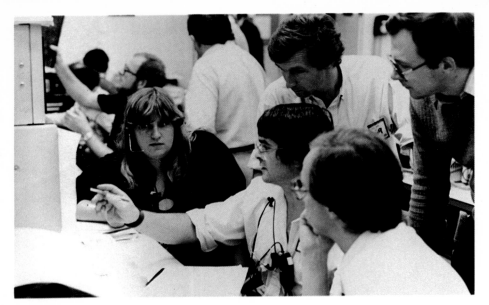
Teams of scientists on the ground worked with the crew in space.

Hundreds of earthbound scientists worked on the problem. Finally one computer was working, and *Columbia* prepared to land. On the afternoon of December 8, 1983, it streaked down from its 155-mile-high orbit.

Spacelab crew leaves the *Columbia* after a successful ten-day mission.

Commander Young brought
Columbia down perfectly
onto the runway at
Edwards Air Force Base in
California. Experts began
repairing the computers for
future flights.

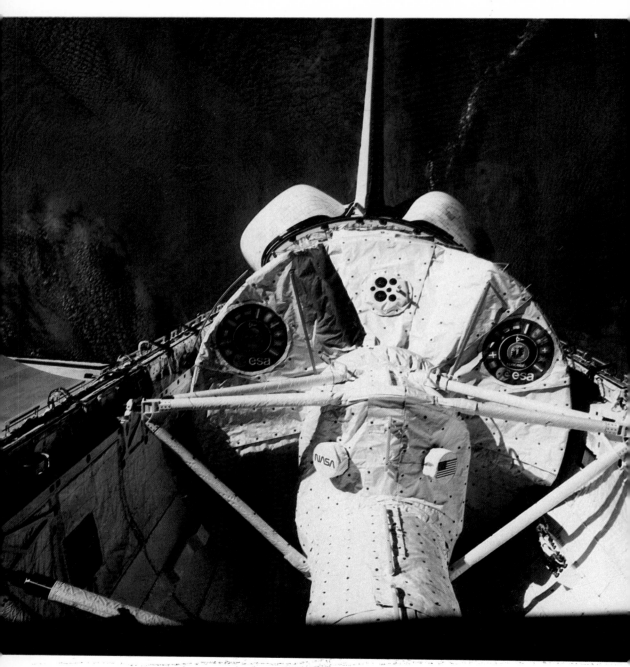

The tunnel in the foreground connects the Spacelab
to the living quarters aboard the *Columbia*.

FUTURE
SPACELAB MISSIONS

Spacelab can be used again and again. It took ten years to build Spacelab and cost $1 billion, so that's an important feature.

More Spacelab missions are planned. Some of the biggest scientific discoveries of the 1980s and 1990s will be made high above earth—in Spacelab!

WORDS YOU SHOULD KNOW

alloy(AL • loi) — a mixture of a metal with a second substance

astronauts(AST • roh • nawts) — American space travelers

astronomy(ast • RON • ah • me) — the study of stars, planets, and other heavenly bodies

atmosphere(AT • mus • feer) — the gases surrounding a heavenly body

circadian rhythms(seer • KAY • dee • yan RIH • thimz) — the daily rhythms experienced by living things

commander(kuh • MAN • der) — the person in charge

deuterium(doo • TAIR • ee • um) — a heavy form of hydrogen

earth(ERTH) — the planet on which we live

experiments(ex • PAIR • ah • mintz) — tests made by scientists to check theories or discover facts

gravity(GRAV • ih • tee) — the force that holds us down to earth

laboratory(LAB • rah • tor • ee) — a place where scientists do experiments

mass-produced(MASS—pro • DOOSED) — made in great amounts

orbit(OR • bit) — the path an object takes when it moves around another object

pallet(PAL • it) — an exposed platform containing Spacelab instruments

reusable(re • YOO • zah • bil) — something that can be used again and again

science(SYE • ence) — the body of knowledge concerned with how and why things are the way they are

shuttle(SHUT • til) — a vehicle that transports cargo or people over a prearranged route

space(SPAISS) — the region beginning about one hundred miles above the earth

Spacelab(SPAISS • lab) — a reusable space laboratory first sent aloft in November of 1983

space station(SPAISS STAY • shun) — an artificial satellite on which people can live and work
star(STAR) — a giant ball of hot, glowing gas
sun(SUN) — the star closest to the earth
telescopes(TEL • eh • skopez) — instruments that make distant objects look closer

INDEX

About the Author

Dennis Fradin attended Northwestern University on a partial creative writing scholarship and graduated in 1967. He has published stories and articles in such places as Ingenue, The Saturday Evening Post, Scholastic, Chicago, Oui, *and* National Humane Review. *His previous books include the Young People's Stories of Our States series for Childrens Press, and* Bad Luck Tony *for Prentice-Hall. In the True book series Dennis has written about astronomy, farming, comets, archaeology, movies, the space lab, explorers, and pioneers. He is married and the father of three children.*